JONES & BARTLETT LEARNING
CDX Automotive

Fundamentals of

Medium/Heavy Duty Diesel Engines

STUDENT WORKBOOK

Gus Wright
Centennial College

JONES & BARTLETT
LEARNING

D1157968

World Headquarters
Jones & Bartlett Learning
5 Wall Street
Burlington, MA 01803
978-443-5000
info@jblearning.com
www.jblearning.com

Jones & Bartlett Learning books and products are available through most bookstores and online booksellers. To contact Jones & Bartlett Learning directly, call 800-832-0034, fax 978-443-8000, or visit our website, www.jblearning.com.

Production Credits
General Manager: Douglas Kaplan
Executive Publisher: Vernon Anthony
Managing Editor—CDX Automotive: Amanda J. Mitchell
Editorial Assistant: Jamie Dinh
Senior Vendor Manager: Tracey McCrea
Marketing Manager: Amanda Banner
Manufacturing and Inventory Control Supervisor: Amy Bacus
Project Management & Composition: Integra Software Services Pvt. Ltd.
Cover Design: Kristin E. Parker
Rights & Media Specialist: Robert Boder
Media Development Editor: Shannon Sheehan
Cover Image: Courtesy of Detroit Diesel Corporation.
Printing and Binding: Edwards Brothers Malloy
Cover Printing: Edwards Brothers Malloy

ISBN-13: 978-1-284-09167-0

6048

Printed in the United States of America
20   10 9 8 7 6 5

Contents

CHAPTER 1 Introduction to Diesel Engines 1
    Chapter Review. 1
    Matching 1
    Multiple Choice. 1
    True/False 2
    Fill in the Blank 3
    Skill Drills 4
    Common Emission System Acronyms 5
    Crossword Puzzle 6
    ASE-Type Questions 7

CHAPTER 2 Careers, Employability Skills,
and Workplace Practices 9
    Chapter Review. 9
    Matching 9
    Multiple Choice. 9
    True/False 11
    Fill in the Blank 11
    Labeling. 12
    Skill Drills 14
    Crossword Puzzle 16
    ASE-Type Questions 17

CHAPTER 3 Safety, Personal Protection
Equipment, and First Aid. 19
    Chapter Review. 19
    Matching 19
    Multiple Choice. 19
    True/False 21
    Fill in the Blank 21
    Labeling. 22
    Skill Drills 26
    Crossword Puzzle 29
    ASE-Type Questions 30

CHAPTER 4 Basic Tools and Lubricants 32
    Chapter Review. 32
    Matching 32
    Multiple Choice. 33
    True/False 35
    Fill in the Blank 35
    Labeling. 36
    Materials Used in the Vehicle Industry . . . 38

    Skill Drills 39
    Crossword Puzzle 42
    ASE-Type Questions 43

CHAPTER 5 Fasteners, Locking Devices,
and Lifting Equipment 45
    Chapter Review. 45
    Matching 45
    Multiple Choice. 45
    True/False 47
    Fill in the Blank 47
    Labeling. 48
    Skill Drills 51
    Crossword Puzzle 54
    ASE-Type Questions 55

CHAPTER 6 Basic Engine Terminology
and Operating Principles 57
    Chapter Review. 57
    Matching 57
    Multiple Choice. 57
    True/False 58
    Fill in the Blank 59
    Labeling. 59
    Skill Drills 61
    Crossword Puzzle 63
    ASE-Type Questions 64

CHAPTER 7 Diesel Engine Emissions 66
    Chapter Review. 66
    Matching 66
    Multiple Choice. 66
    True/False 67
    Fill in the Blank 68
    Crossword Puzzle 69
    ASE-Type Questions 70

CHAPTER 8 Diesel Engine Combustion
Systems. 72
    Chapter Review. 72
    Matching 72
    Multiple Choice. 72
    True/False 73